Charlevoix Stones

and Other Beachcomber Treasures
of Northern Michigan

by

Vic Eichler, Ph.D.

Charlevoix Stones
and Other Beachcomber Treasures of Northern Michigan

Copyright © 2012 by Victor B. Eichler

ISBN: 978-0-9703620-6-3

Shantimira Press
P.O. Box 171
Three Rivers, MI 49093-0171
shantimirapress@yahoo.com

Cover design by Julie Taylor.

Printed in the United States of America.

Cover Photo: A group of Charlevoix Stones showing the tiny 'honeycomb' pattern. These stones have been wiped with oil to enhance their appearance.

For Ruth

*Devoted Partner, Dedicated Supporter, and Willing
Walker of Rocky Beaches to satisfy my whim
for finding prize stones.*

Contents

A number of Charlevoix Stones lying among
other stones on a Little Traverse Bay beach

Introduction

I first heard the term "Charlevoix Stone" at one of the early Petoskey Stone Festivals held since 2006 in Barnes Park in Antrim County near the village of Eastport, Michigan. I was there as a vendor, selling my popular book for children, *Peter And His Pet Petoskey Stone*, which glorified the fossilized coral found uniquely in our state.

The rock that has become known as Charlevoix Stone was not unknown to me; it is found quite frequently along the shore of Little Traverse Bay in the northern part of the lower peninsula of Michigan along with an occasional Petoskey Stone and a mix of numerous other rock fragments. My immediate thought was, "Is this a name created by merchants in the city of Charlevoix because they were 'beat out' by the Governor of Michigan having named the official state stone after the nearby city of Petoskey?"

I have made numerous inquiries of merchants and residents of Charlevoix in regard to how the name became attached to a specific coral fossil and regret that the answer appears to be lost in time. Nevertheless, the name Charlevoix Stone is widespread in the area, and it is hoped that this small volume will not only extend the popularity of its type but also distinguish it from other fossilized corals found on the beaches of northwest Michigan.

Let's begin with a look at Charlevoix Stones—both as they may be found "in the rough" along the shore of beautiful Lake Michigan and how they may appear when carefully polished. We will then consider how Charlevoix Stones may be distinguished from Petoskey Stones and other corals found nearby whose skeletons created a variety of fossils. Lastly, we will consider the natural history of the many different organisms that once lived in the warm, shallow sea that covered most of the state of Michigan a long, long time ago and have left their imprint as fossils.

What Does A Charlevoix Stone Look Like?

No doubt Charlevoix Stones and Petoskey Stones are the best known fossils in Michigan, and both are actually the preserved skeletons of coral animals that lived in colonies during the Devonian times, more than 300 million years ago. That was a time when warm, shallow ocean waters covered most of the present boundaries of the state. The little animals lived in close proximity to each other, forming a colony which often created reefs several hundred feet in length.

Fig. 1. Close-up of a "rough" Charlevoix Stone as found on a beach.

When pieces of the colony break off, they may spend hundreds or thousands of years rolling in the tides and become very round and smooth. In appearance they have a honeycomb appearance with very tiny cells packed together. On the next page is an example of one such stone that has been smoothed in this way.

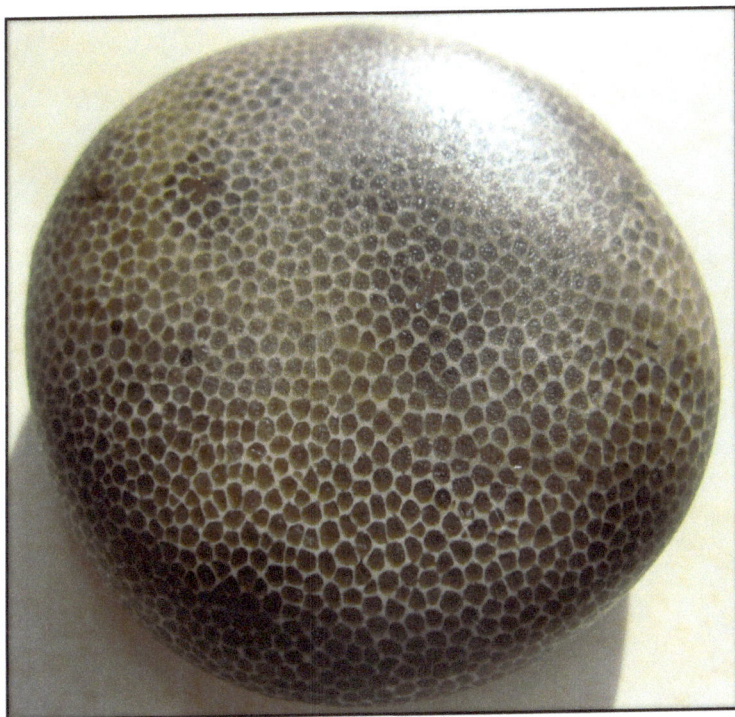

Fig. 2. A Charlevoix Stone, smoothed from action of waves, and oiled to bring out surface detail.

Because of its appearance, it is sometimes called "Honeycomb coral." Some people have referred to Charlevoix Stones as 'miniature' or 'baby' Petoskey Stones, which is clearly incorrect. The two stones are comprised of different species of coral. Each one of the "eyes" seen on specimens represent where the small, soft bodied coral animal (called a 'polyp') lived. The coral polyps of the genus *Favosites* (of which all species are now extinct) that formed the Charlevoix Stones were just much smaller than individual corals of the Petoskey Stone.

How Are Charlevoix Stones Different From Petoskey Stones?

Below is a Petoskey Stone of about the same size as the Charlevoix Stone in Fig. 2. Note how large the mineralized chambers are in comparison. This fossil coral is in a different genus than the Charlevoix Stone, *Hexogonaria,* which means that the individuals are roughly hexagons, or six-sided figures. Hexagons are often found in nature because individual elements can be efficiently packed together. The wax comb of honeybees is a good example. This colony of corals possibly was living when the first vertebrate animals that were to give rise to amphibians ventured onto the land about 360 million years ago.

Fig. 3. A Petoskey Stone has much larger individual polyps than the Charlevoix Stone.

It should be mentioned that on June 28, 1965, Michigan's Governor George Romney signed a bill that made the Petoskey Stone the official stone for the state. Michigan thus became the first state to recognize a fossil as the Official State Stone!

How Do Corals Grow?

The individual coral animal is closely related to sea anemones, and in both forms the individual attaches to a substrate during development. They are distinguished, however, in that the polyp of sea anemones generally develop as individuals while the polyps of corals develop as colonies. The coral polyps reproduce by the process known as 'budding,' in which offspring are formed from the mature parent's body and remain attached, thus being capable of forming reefs as the population increases over time.

The body of the polyp is like a two-layered sac with a gelatinous material between the outer and inner layers. Coral animals create a hard mineralized skeleton around the soft tissues of their bodies and are therefore unable to relocate once the skeleton has formed. A circle of tentacles surrounds the central mouth at the free, unattached end. The mouth of the polyp is surrounded by tentacles used to capture food. Stinging structures called nematocysts, located on the tentacles, are capable of firing dart-like structures that contain toxins capable of paralyzing prey that the polyp feeds on.

Fig. 4. Sketch of coral polyps with tentacles extended.

A longitudinal section through a colony may look like a bunch of long soda straws bundled together, as can be seen at the right side of the photo below.

Fig. 5. This slice through a Charlevoix Stone shows a cross-section of the small chambers that the coral animals lived in (center and left) and the longitudinal nature of the chambers (at right) as they extend toward the point where they were anchored.

The reefs that corals collectively form are still found in warm tropical seas worldwide. They provide a home for many thousands of different fish and other marine animals, providing places for breeding and feeding. The main chemical comprising coral reefs is calcium carbonate secreted by the soft bodies. Coral reefs are extremely sensitive to changing environmental conditions, especially pollution and physical damage caused by human activities, and are in great danger of being destroyed. To emphasize the importance of the reefs, they cover less than 1% of the ocean floors but are important to 25% of the plant and animal species found in the oceans.

Why Are These Fossils Particularly Found In This Area?

The time when the coral reefs from which genus *Favocites* and *Hexogonaria* species flourished—those which gave rise to the Charlevoix Stones and Petoskey Stones—was in the Middle Devonian Period which dates back to approximately 350 million years ago. As the calcium carbonate mineral skeletons of the coral polyps fossilized, they became incorporated into sedimentary rock layers that were also forming at that time. Known as the Alpena Limestone Formation, this fossil-bearing rock lies deep underground, except it reaches the surface in the area around Little Traverse Bay in the northern point of Michigan's lower peninsula. Both of the cities of Charlevoix and Petoskey are located in this area, and hence the rocks which were created from the beautiful fossil coral skeletons were named for these two cities.

During the last ice age, approximately 10,000 years ago, glaciers that extended over the land in several lobes from the north carried some pieces of the Alpena Formation rock to the south. In the process the pieces broke into fragments that now are usually only a few inches wide.

The distribution of the stones is limited to the regions where the glaciers advanced and then retreated, which is quite limited to the lower peninsula of Michigan, the shores of Lake Michigan and inland gravel pits. The farther south one travels in the state, however, the rarer is the likelihood that either of these special stones will be found.

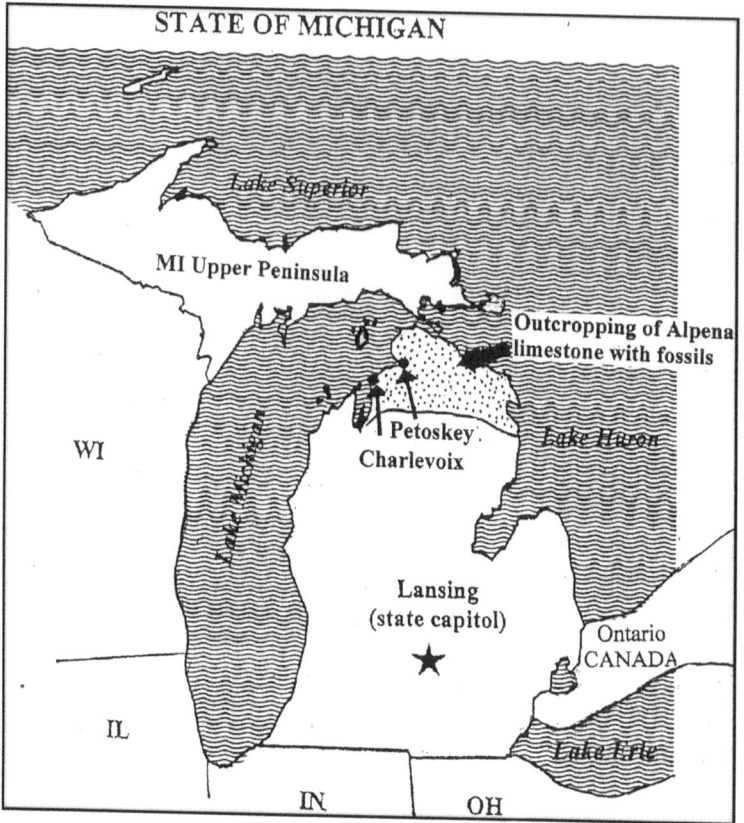

Fig. 6. A map of the state of Michigan showing the region of the Alpena Limestone Formation (stippled area) where Charlevoix Stones and Petoskey Stones originate. Location of the cities of Petoskey and Charlevoix are also shown.

How Did These Fossils Form As Stones?

A fossil is a remnant of a plant or animal that lived millions of years ago and has been turned to stone by a process of mineralization and has become preserved. Fossils give us clues to what life was like in the remote past.

Some fossils are spectacular, such as the preserved bones of dinosaurs, large animals which have been extinct for many tens of millions of years. Others may be microscopic, such as the pollen grains from a tree that lived in the remote past. Soft tissues generally do not preserve well as fossils, so hard structures such as bones, teeth and shells most often are found as fossils.

Preserved footprints of ancient reptiles and amphibians that lived in swamps are considered to be fossils, as are impressions of leaves of trees that have been preserved when encapsulated in mud that has become hard as stone.

Scientists who study the Earth's history have divided the 4.5 billion years since the planet was formed into four divisions called EONs. Life was extremely primitive in the most distant eons, and only the most recent, called the Phanerozoic (or 'visible life'), concern our study of corals. This most recent period is divided into three ERAs, and these are described on the next page. Geologically, the oldest rocks, and oldest fossils when present, appear lower in the rock layers, and rocks formed more recently are closer to the surface of the land.

The Eras recognized by scientists in the Phanerozoic Eon are:

Paleozoic ('old life') began about 540 mya (million years ago) and extended to 248 mya. During the almost 300 million years of the Paleozoic Era, the first corals appeared around 430 mya. During this Era primitive plants first appeared on land and became very diverse, eventually forming extensive forest swamps which created most of the world's coal deposits as the vegetation decayed and became compressed. A variety of fish were dominant in the seas during the Devonian Period (412-354 mya) of this era. During later periods of the Paleozoic, amphibians and primitive reptiles began colonizing the land.

Mesozoic ('middle life') spans 183 million years, beginning 248 mya and extended to 65 mya. During the Mesozoic ferns flourished and flowering plants first appeared on the land. The "Age of Reptiles" began as they became dominant forms on the land, in the waters, and in the air. By the end of this era, however, large dinosaurs had become extinct.

Cenozoic ('recent life') began 65 mya and extends to the present time. Flowering plants became widespread, large flightless birds with true feathers were present early in the Cenozoic and evolved to modern forms. Populations of small mammals which survived from the Mesozoic expanded to colonize the land, and the "Age of Mammals" began. Small horses, rhinoceroses, camels, and other large mammals became present, but widespread extinction in the latter periods of the Cenozoic Era left fossils of such mammals as sabre-toothed cats, cave bears, and woolly mammoths. Primates, including early humans, became present late in the Cenozoic, beginning about 5 mya and can be traced through successive periods and epochs of the geological record to modern forms.

What Other Interesting Fossils and Stones Might Be Found On The Shore?

Less well known are the fossils created from other species of coral animals that are often found in the same vicinity as Charlevoix Stones. Among beach rocks might be seen any of the following:

A. More Coral Fossils That You Might Find

1. Chain Coral

The attractive fossil known as "Chain Coral" may be found along with Charlevoix and Petoskey Stones, but they are not abundant. When living, they belonged to the family of tabulate corals known as Halysitidae and the genus *Halysites*. This genus of coral is known to have been present for about 45 million years but became extinct about 416 million years ago.

Fig. 7. A group of Chain Coral showing typical rows of coral skeletons as links in a chain.

The individual coral polyps were found in colonies where they constructed their individual skeleton much like an elliptical tube in linear fashion. As the colonies grew, these tubes took the appearance of links in a chain when viewed from the open ends. These corals grew in the warm, shallow tropical seas and contributed to the reefs of the period.

2. Horn Coral

This unusual and attractive coral is identified by its conical shape, much like a bull's horn. These were solitary animals that attached to the sea bottom at the small end and had tentacles by which they caught food extending from the expanded end. They are often marked by lines or ridges running from small to large end.

Fig. 8. One horn coral as found (left) and two polished and buffed horn corals (right).

B. Fossils Found On The Beach That Are Not Corals

1. Brachiopods

Brachiopods are a large group of marine invertebrates with two shells that are hinged to allow the soft-bodied animal within to open to obtain food, to shed reproductive cells, and to eliminate wastes. There are about 300 living species of brachiopods, most under 4" long. Many of the 10,000+ species of brachiopods known from fossils grew to twice that size.

Fig. 9. A fossil impression of one shell of a brachiopod.

2. Crinoid Stems

Among the fossils of echinoderms (which include starfish and sea urchins) are the crinoids—organisms whose modern types look perhaps more like a beautiful flower growing from the sea floor than an animal. Crinoid fossils are found worldwide, for these animals were a dominant form during the Paleozoic Era.

The main body of the crinoid was a stem that was often several meters long ending with flowing tentacles that gave them the nickname 'sea lilies.' The stems consisted of segments that are often found singly or in columns. Often times, large numbers of the individual disks are found in limestone and are either solid, like coins, or round with a hole within like a bead.

Fig. 10. Left Photo: A rock comprised of petrified ocean bottom mud in which sections of crinoids are embedded. Right Photo: A long piece of a crinoid stem.

C. Other Interesting Beach Stones

1. Conglomerates

Conglomerates are sedimentary rocks in which rock pieces of various size are cemented along with sand and pebbles with dissolved minerals. Heat and pressure during long periods of geological time add to the firmness in which the mix is held together. The pebbles and small rocks in a conglomerate are typically rounded, a feature that differentiates them from breccias in which the larger stones in the mix are angular rather than round.

Fig. 11. An example of a conglomerate group of pebbles and sand cemented by minerals, heat, and pressure over time.

2. Pudding Stones

Pudding Stones are conglomerates that formed from glacial debris in Canada and were brought south notably to the region of Drummond Island off the east coast of Michigan's upper peninsula. The base rock is fine grained and light colored quartzite matrix but has inclusions within that contrast markedly with the surrounding color. Typically Michigan Pudding Stones have inclusions of two minerals present: reddish jasper and black hematite. The name was given to this stone by British settlers in the 1800s who thought that the stone looked like the raisins and cherries found within puddings in their native England.

As glaciers advanced over the state from the north, Pudding Stones were pushed or carried with the ice and hence may be found throughout the state of Michigan.

Fig. 12. Two small Pudding Stones as found (left),
and a cut and polished section (right).

3. Michigan Septarians

The interesting stone found on Michigan shores that appears like a brown stone with whiteish partitions is called a Michigan septarian (in distinction from other appearing septarians found elsewhere in the world). The name comes from the root word 'septa', meaning 'partitions'. These rocks form from mud at the sea bottom that cracks as it dries. Subsequently, calcite or other minerals fill in the cracks to give the unique appearance.

Fig. 13. A group of Michigan Septarians as found on the beach

4. "Stinkstones"

These rocks are very common on many Lake Michigan beaches. They are light colored and punctuated with numerous small holes. These stones contain hydrogen sulfide gas—the same gas that gives rotten eggs their unmistakable odor. If stinkstones are smashed against a larger rock and smelled immediately, your nose will no doubt confirm that you have correctly identified this stone.

Fig. 14. Photo of a Stinkstone, characterized by numerous small pores

Notes

About the Author

From his early years, Vic Eichler has had a deep connection to the natural world. An avid collector and classifier, he continues to love the adventure of exploring various habitats and learning their elements and mysteries.

As a retired professor of Biological Sciences, Vic has time and a desire to share his fascination and knowledge of the out-of-doors with children and adults.

Since the term "Nature Deficit Disorder" has become popular to recognize how electronic devices and indoor recreation now occupy time that people might have formerly spent outside the home, he feels that providing concise, illustrated, and inexpensive guides to nature are a contribution to making our outdoor surroundings more relevant and understandable to his readers.

Tourism is one of the most important industries in the state of Michigan, and with its extensive coastline on four of the five Great Lakes, time spent at a beach is a popular pastime for families.

While this guide was specifically created for the coast of Lake Michigan along the northern region of the state, many of the stones and fossils pictured here may be found in other regions of the state.

If you are—or want to be—a beachcomber looking for the perfect stone, I hope that this book is helpful, and that you enjoy the hunt!

Other Books by Vic Eichler

Published by Shantimira Press

Morel Mushrooms in Michigan and Other Great Lakes States
A guide to where, when and how to find morels; How to pick, preserve, cook and enjoy morel mushrooms. Includes numerous gourmet mushroom recipes from a celebrated Michigan chef.
40 pages; $11.95

Peter and His Pet Petoskey Stone
A story and guide for finding and polishing Michigan's official state stone.
24 pages; $5.95

Emily Joy & The Woman Who Trained Horses
A glimpse into the fascinating life of Madame Marantette of Mendon, Michigan.
52 pages; $14.95 (currently out of print)

The Awakening of Freddy Tadpole
A story not only of amphibian metamorphosis, but the transformation of one being to surrender fear of the unknown. A story for 'seekers' of all ages.
48 pages; $14.95

People in the Neighborhood
A delightful collection of published articles about interesting, individualistic, and often eccentric people living, working and playing in St. Joseph County, Michigan. Large format (18" x 11").
48 pages; $18.95

www.ingramcontent.com/pod-product-compliance
Lightning Source LLC
Chambersburg PA
CBHW041224270326
41933CB00001B/34